Contents

Preface...3

About the Author...4

Chapter 1: The Living World...5-21

Biodiversity

Types of Biodiversity

Three Domains of Life

Taxonomy

Systematics

Concept of Species

Taxonomical Hierarchy

Binomial Nomenclature

Museums

Zoological Parks

Herbaria

Botanical Garden

Chapter 37: Biodiversity and It's Conservation..22-41

Biodiversity Patterns

Importance

Loss of Biodiversity

Biodiversity Conservation

Hotspots

Endangered Species

Red Data Book

Biosphere Reserves

National Parks

Sanctuaries

Ramsar Sites

Preface

Welcome to biodiversity. Biodiversity is the variety and variability of life on Earth. Without biodiversity, we do not live. Human population is increasing day by day and destroy the biodiversity for various needed. Biodiversity refers to varieties of living things i.e plants, animals, bacteria, and fungi. Biodiversity provides food, water, timber, fiber, and genetic resources.

The book comprises two chapters. The first chapter gives definition of living, biodiversity, taxonomy, systematics, and species. The first chapter also gives types of biodiversity, three domains of life, taxonomical hierarchy, binomial nomenclature, role of botanical garden, importance of herbaria. The second chapter deals with the biodiversity patterns, importance of biodiversity, and loss of biodiversity. The second chapter also discusses biodiversity conservation, hotspots, endangered species, red data book, biosphere reserves, national park, sanctuaries, and ramsar sites.

Thanks to my family. Finally, thanks to all the friends for their continue support and encouragement.

Preface

Welcome to biodiversity. Biodiversity is the variety and variability of life on Earth. Without biodiversity, we do not live. Human population is increasing day by day and destroy the biodiversity for various needed. Biodiversity refers to varieties of living things i.e plants, animals, bacteria, and fungi. Biodiversity provides food, water, timber, fiber, and genetic resources.

The book comprises two chapters. The first chapter gives definition of living, biodiversity, taxonomy, systematics, and species. The first chapter also gives types of biodiversity, three domains of life, taxonomical hierarchy, binomial nomenclature, role of botanical garden, importance of herbaria. The second chapter deals with the biodiversity patterns, importance of biodiversity, and loss of biodiversity. The second chapter also discusses biodiversity conservation, hotspots, endangered species, red data book, biosphere reserves, national park, sanctuaries, and ramsar sites.

Thanks to my family. Finally, thanks to all the friends for their continue support and encouragement.

Chapter 1

The Living World

What is living?

Living means alive. Living things are made up of cells, that requires energy. Living things can grow, move, reproduce, and respond to their environment.

Biodiversity:

Biodiversity is the variety and variability of life on Earth. Without biodiversity, we do not live. Human population is increasing day by day and destroy the biodiversity for various needed. Biodiversity refers to varieties of living things i.e plants, animals, bacteria, and fungi. Biodiversity provides food, water, timber, fiber, and genetic resources.

According to U.S. Office of Technology Assessment (1987), biological diversity is "the variety and variability among living organisms and the ecological complexes in which they occur."

Types of Biodiversity:

There are three types of biodiversity-

i. Genetic Diversity: Genetic diversity is refers to the diversity in the genetic makeup within the species.

ii. Species Diversity: Species diversity is the number of species live in a particular location. Species diversity refers to the number of species found in ecological community.

iii. Ecosystem Diversity: Ecosystem diversity is the variety of ecosystem within an area. Ecosystem diversity includes both biotic and abiotic component.

Three Domains of Life:

The evolutionary relationships between organisms are subject of phylogeny. The phylogenetic tree showing the evolutionary relationships among various biological species or other entities. All life on earth is part of a single phylogenetic tree, indicating common ancestry. The microbial world has three main cell lineages which are thought to have evolved from a single progenitor. The lineages are formally knowm as Domain. The three domain system, proposed by woese and others, is an evolutionary model of phylogeny based on differences in the sequences of nucleotides in the cell's rRNA, as well as cell's membrane lipid structure and its sensitivity to antibiotics. The phylogenetic tree consist of three domains of organisms the Bacteria and the Archae, cells of which are prokaryotic, and the Eukarya (eukaryotes). Eukaryota and Archaea are more closely related to each other than Bacteria (based on Caliver-Smith's theory of bacterial evolution). Each of these three domains contains unique rRNA. rRNA is the RNA component of the ribosome, which are essential for protein synthesis in all living organisms.

Property	Archaea	Bacteria	Eukarya
Cell membrane	Glycerol diether linked lipids.	Ester-linked phospholipids and hopanoids.	Ester-linked phospholipids and sterols.
Cell wall	Glycoproteins	Peptidoglycan	Various structure
Endoplasmic reticulum	Absent	Absent	Present
Golgi apparatus	Absent	Absent	Present
Lysosome	Absent	Absent	Present
Mitochondria	Absent	Absent	Present
Chloroplast	Absent	Absent	Present
Nucleolus	Absent	Absent	Present
RNA polymerase	Many	One	Many
Toxin	Sensitive to diphtheria toxin	Resistant to diphtheria toxin	Sensitive to diphtheria toxin
Histone	Absent	Present	Present

Taxonomy:

Taxonomy is come from two greek words taxis meaning arrangement and nomia meaning method. Taxonomy encompasses the description, identification, nomenclature, and classification. Taxonomy includes all plants, animals, and microorganisms.

The term taxonomy was first coined by A.P de Candolle in 1813.

Systematics:

The term taxonomy and systematics have been interchangeably used in the past. According to Radford in 1986, systematics is the study of phenotypic, genetic, and phylogenetic relationship among taxa.

Objectives of Systematics:

i. Understading of the evolutionary process.

ii. To establish a suitable method for identification, nomenclature, and description of plant taxa.

Concept of Species:

Species: Species is refers to the group of living organisms that can be produce fertile offsprings.

Taxonomical Hierarchy:

Taxonomical Hierarchy is the sequence of arranging various organisms into the biological classifications.

Here, Kingdom is the highest rank and species is the lowest rank in this hierarchy.

Binomial Nomenclature:

Binomial nomenclature is the system of naming species or organisms of living things.

All living organisms like plants, animals, birds, and some microorganisms have their own scientific names. For e.g.

- The Scientific name of the Mango is represented as Mangiferaindica. Here, Mangifera represents the genus and indica represents the particular species.
- The Scientific name of the Leopard is presented as Pantherapardus . Here Panthera represents the genus and pardus represents a species.

Codes:

ICZN- International Code of Zoological Nomenclature.

ICNafp- International Code of Nomenclature for algae, fungi, and plants.

ICNB- International Code of Nomenclature of Bacteria.

ICTV- International Committe on Taxonomy of viruses.

Botanical Gardens:

Botanical Gardens are the institutions. Botanical Gardens maintain the collection, preservation, and cultivation of different varieties of plants.

Role of Botanical Garden:

i. Botanical Garden provide information of local flora.

ii. Botanical Garden supply seeds, flowers, fruits, and materials.

iii. Many peoples are visit botanical garden. Botanical gardens attract many peoples.

iv. Botanical Garden conserve rare species.

v. Botanical Garden provide education to teacher, students, and scientists.

List of botanical gardens in India:

i. Indian Botanic Garden, Kolkata.

ii. Botanical Garden, Sharangpur.

iii. Lalbagh Botanical Garden, Bangalore.

iv. Government Botanical Gardens, Ooty.

v. Tropical Botanic Garden, Trivandrum.

vi. Lloyd Botanic Garden, Darjeeling.

vii. Jawaharlal Nehru Botanical Garden, Gangtok.

viii. National Botanic Garden, Lucknow.

ix. Assam State Botanical Garden, Guwahati.

x. SemmozhiPoonga Botanical Garden, Chennai.

xi. Pilikula Botanical Garden, Mangalore.

xii. Reddy Botanical Garden, Hyderabad.

xiii. Botanical Garden, Chandigarh.

xiv. Waghai Botanical Garden, Saputara.

xv. Nehru Memorial Botanical Garden, Srinagar.

Table 1. List of Botanical Gardens in various country:

Country	Botanical Gardens	Formed/Founded
Abkhazia	Sukhumi Botanical Garden	1840
Argentina	Administraction de parquesNacionales	September 30, 1934
Argentina	Buenos Aires Botanical Garden	September 7, 1898
Armenia	Yerevan Botanical Garden	1935
Armenia	IjevanDendropark	1962
Armenia	Sevan Botanical Garden	1944
Armenia	StepanavanDendropark	1931
Austrilia	Austrilian National Botanic Garden	1949
Austrilia	Lindsay Pryor National Arboretum	
Austrilia	Westbourne Woods	
Austrilia	Albury Botanic Gardens	1877
Austrilia	Auburn Botanical Garden	1977
Austrilia	Austrilian Inland Botanical Garden	1989
Austria	Innsbruck University Botanic Garden	1911
Austria	Botanical Garden of the University of Vienna	
Bangladesh	National Botanical Garden of Bangladesh	1961
Bangladesh	Balda Garden	1909
Barbados	Andromeda Botanical	

	Gardens	
Barbados	Hunte's Gardens	
Belgium	Antwerp Botanic Garden	1825
Belgium	Botanical Garden of Brussels	1826
Belgium	Arboretum Kalmthout	1856
Belgium	HortusBotanicusLovaniensis	1837
Belize	Belize Botanic Gardens	
Bermuda	Bermuda Botanical GardenS	
Botswana	National Botanical Garden	November 2, 2007
Singapore	Singapore Botanic Gardens	1859
Singapore	Gardens by the Bay	29 June 2012
Vietnam	Saigon zoo and Botanical Garden	1865
United States	Alaska Botanical Garden	1993
United States	Georgeson Botanical Garden	
Uganda	National Botanical Gardens	1898
Zimbabwe	National Botanical Garden of Zimbabwe	1902
India	Sanjay Gandhi JaivikUdyan, Patna	1973
India	National Cactus and Succulent Botanical Garden and Research Centre, Haryana	1987

Herbarium:

Herbarium is the collection of plant materials in any place. After collection of plants, plants should be dried, pressed and arranged according to the classifications system. Herbarium may be mounted on a sheet of paper.

Luca Ghini is the creator of the art of herbarium.

Figure 4. Herbarium specimens of Crotalaria juncea (Photo Credit: AnupamRajak)

Importance of Herbaria:

i. Herbaria are used in taxonomic research.

ii. Provide data for floristic studies.

iii. Herbarium provide knowledge about the flora.

Table 1. List of Herbaria:

Name	No. Specimens
Museum of Natural History, Paris	6.5 million
Komarov Botanical Institute, Leningrad	Over 5 million
Conservatory and Botanical Garden, Geneva	5 million
Combined Herbaria, Harvard University, Cambridge	4.5 million
New York Botanical Garden, Bronx	4.3 million
U.S National Herbarium	4.1 million
British Museum of Natural History, London	4 million

Museum:

Museums are the institution where educational materials are showing to the public. Museums preserve selected objected. Museums give education to the public.

Figure 5. Museums (Photo Credit: Pixabay)

Table 1. List of Museums

Country	Name of the Museum	Established

India	Bhagwan Mahabir Government Museum	1982
India	Victoria Jubilee Museum	1887
India	Assam State Museum	1940
India	Goa Chitra Museum	2010
India	Swaminarayan Museum	2011
United States	Alabama Administrative office of Courts Museum Area	1994

Zoological Park:

Zoological parks are the institution where living organisms are kept and exhibited to the public.

List of Zoological parks-

i. National Zoological Parks, New Delhi.

ii. Rajiv Gandhi Zoological Park, Maharashtra.

iii. Nandankanan Zoological Park, Odisha.

iv. Indira Gandhi Zoological Park, Andhra Pradesh.

v. Padmaja Naidu Himalayan Zoological Park, Darjeeling.

Chapter 2

Biological Classifications

Five Kingdom Classification:

The five kingdom classifications system was developed by Robert H. Whittaker in 1969.

They classified the five major groups-

- Kingdom- Animalia.
- Kingdom- Plantae.
- Kingdom- Protista.
- Kingdom- Monera.
- Kingdom- Fungi.

Kingdom- Animalia:

i. Animalia is also called as Metazoa.

ii. Animalia are multicellular, and heterotrophs.

iii. Animalia lack cell wall.

iv. Animals are eukaryotic.

v. They are composed of collagen and glycoproteins.

Kingdom- Plantae:

i. They are eukaryotic.

ii. Plants are contains cell wall. Cell wall is made up of cellulose.

iii. They contains plastids and chlorophyll.

iv. They store food as starch.

v. They reproduce sexually or asexually.

Kingdom- Protista:

i. They are eukaryotic.

ii. They can be unicellular or multicellular.

iii. They reproduce sexually or asexually.

Kingdom- Monera:

i. Monera is a prokaryotic.

ii. They are unicellular organisms which contains 70S Ribosomes.

iii. They lack cell organelles like mitochondria, lysosomes, plastids, golgi bodies, endoplasmic reticulum, centrosome etc.

Kingdom-Fungi:

i. Fungi are found in dump places.

ii. Fungal cell wall is composed of glucans and chitins.

iii. Fungi reproduce both sexually or asexually.

Classifications of Monera:

Monera is classified into three sub kingdom-

i. Archaebacteria.

ii. eubacteria.

and, iii. Cyanobacteria.

Archaebacteria:

Archaea is derived from the Greek word archaios, meanning ancient. Archaea are found all over the world. In 1977, Carl Woese and his colleagues including George Fox, while working at the University of Illinois discovered the domain archaea. Archaea are the ancestors of all eukaryotes because archaea have more common on similarities with eukaryotes and bacteria. So, many articles and book refer to them as archaebacteria. Generally, archebacteria can produce methanes.

Archaebacteria can live in different habitats. Some live extreme alkaline or acidic and hot ($100°C$) temperature. They have been also found inside the human gut, digestive tracts of cow, termites where they produce methane. Researcher also suggests that archebacteria are also found in the plankton of the open sea. They were originally discovered in hydrothermal vents and terrestrial hot spring.

Characters of the Archaea

Archaebacteria have number of characteristics:

1. The individual cells are prokaryotic in nature which contains no nucleus.

2. Archaea lack of peptidoglycan in their cell wall.

3. Achaea may be spherical, rod shaped, rectangular or irregular in shape and pleomorphic in nature.

4. Like bacteria, archaea are unicellular organisms.

5. Archaebacteria use pigment bacteriorhodopsin (light energy transduction-pigment) for photosynthesis. So, they are autotroph.

Archaea Domain

Archaea are classified into three domain. Under the archaea domain, they are:

i. Crenarchaeota.

ii. Euryarchaeota.

iii. Korarchaeota.

i. Crenarchaeota : The kingdom Crenarchaeota are microscopic, unicellular, grow between 80°C and 100°C.

ii. Euryarchaeota: Euryarchaeota organisms are found in salt lake, sea water, and hypersaline environment. Pyrococcusfuriosus (Pfu) is an extremely thermophiliceuryarchaeota used in the PCR reaction.

iii. Korarchaeota: Korarchaeota organisms are grows in 70-97°C.

Eubacteria:

Bacteriology is the science that deals with the study of organisms known as bacteria (singular, bacterium). The word germ is probably synnomous with bacterium. Bacteria among the simplest form of life know and hence show characteristic of both plants and animals. Bacteria are small (microscopic size) organisms that can be found in most environments, for example in soil, water, and on and inside the human body. There are around 50 million bacteria in every gram of surface soil. Bacteria can be harmful, but some species of bacteria are needed to be healthy. Research suggests that efforts to make a cleaner environment, free from bacteria, are contributing to the rise in obesity, cancer and heart disease.

Size

Bacteria are considerably smaller than yeasts, moulds, algae, and protozoa. They vary greatly in size according to species.

Bacteria show considerable variation in size according to species. Regardless of their size none are visible with naked eye. Some bacteria are large enough to see with naked eye. Whereas Schaudinnumbutschlii that measure between 4 and 5 μm in diameter are considered to large bacteria.

Table 3.1 Size of bacteria

Bacteria	Size (μm)
Cocci	0.5μm to 1.25 μm
Bacillus or rod shaped bacteria	0.5 μm -1.0 μm X 2-3 μm
Helical or spiral bacteria	1.5 μm
Euplopisciumfishelsohni	200 μm X 80 μm
Thiomargaritanamibiensis	750 μm

5.1 Morphology (shape) of bacterial cell

The three basic bacterial shapes are coccus (spherical cell), bacillus (rod shaped), and spiral (twisted forms). Cocci are bacteria that are spherical, ovoid, or generally round in shape. Cocci occur in single cell or remain attached, that can be based on cellular arrangement. Cocci are of different types:

• Diplococci – Occur in pairs. e.g Streptococcus pneumonia and Neisseria gonorrhoeae, Moraxella catarrhalis.

• Styphalococci – Irregular (graph lke) cluster of cocci. e.gStyphalococcusaureus.

• Streptococci – Occur in long chain. e.g Streptococcus lactis, Streptococcus pyogene.

• Tetracocci – Tetrads are square arrangement of four cocci. e.gAerococcus, Pediococcus and Tetragencoccus.

- Sarcinae – The cocci are arranged in a cuboidal manner made up of 8 or more cells. e.gSarcinaeventriculi, Sarcinaeureae etc.

Arrangement of baccilus

The cylindrical or rod shaped bacteria are called bacillus. They may be motile or non motile.

- Diplobacilli – Occur in pair. e.g bacillus cereus.

- Streptobacilli – Occur in chain. e.g Bacillus moniliformis.

Spiral bacteria

Spiral are curved bacteria which can range from a gently curved shaped. Many spirili are rigid and capable of movement.

Besides the above groups, following other shapes of bacteria are also present –

- Vibrio – They are comma shaped bacteria with less than one complete turn or twist in the cell. e.g Vibrio chlorea.

- Filamentous – Some bacteria are filament like. e.gBeggiatoa, Thiothrix.

- Spirilla – They have a rigid spiral structure. e.g Helicobacter pylori, Spirillumwinogradskyi.

- Pleomorphic – Some bacteria are able to change their shape and size in response to variation in the surrounding environment. In pure culture, they can be observed to different shapes.

Structure of bacterial cell wall

The cell wall is usally fairly rigid just outside the plasma membrane. A bacterial cells show typical prokaryotic structure. Species of bacteria can be divided into two major groups, called gram positive and gram negative. The distinction between gram - positive and gram - negative based on the gram stain reaction. Gram positive bacteria stained purple whereas gram negative bacteria were coloured, pink or red. Bacterial cell wall composed of: i. Peptidoglycan. Ii. Outer membrane and iii. Surface membrane.

Peptidoglycan

Peptidoglycan is a polymer composed of two sugar derivatives – N-acetylglucosamine and N- acetylmuramic acid and a few amino acids, including L-alanine, D-alanine, D-glutamic acid and either lysine or diaminopimelic acid (DAP). The sugar component consists of alternating residues of beta (1,4) linked N – acetylglucosamine (NAG) and N – acetylmuramic acid. The peptidoglycan layer is thicker in Gram positive bacteria (20-80 nanometers) then in Gram negative bacteria (7-8 nanometers). The gram positive bacteria have substances called teichoic and teichuronic acids interspersed with the peptidoglycan polymer.

Table 5.1 Comparision of the cell walls of Gram-positive and Gram-negative bacteria

Gram-positive cell wall	Gram-negative cell wall
1. Thick peptidoglycan layer.	1. Thin peptidoglycan layer.
2. Principal surface antigen is teichoic acid.	2. Principak surface antigen is lipopolysaccharides.
3. Cell wall is rigid.	3. Cell wall is elastic.

4. Cell wall is single layered.	4. Cell wall is multilayered.
5. Teichuronic acid is present.	5. teichuronic acid is absent.
6. Presence of glycolipids.	6. Absence of glycolipids.
7. Some examples of gram-positve bacteria are Streptococcus, Clostridium, Lactobacillus etc	7. Some examples of gram negative bacteria are Vibrio, Rhizobium, Escherichia coli, Acetobacter.

Cyanobacteria:

i. Cyanobacteria are unicellular.

ii. Cyanobacterial cells are prokaryotic in nature.

iii. Cyanobacteria are also known as blue green algae. They are found in water.

iv. They are photosynthetic.

Classification of Protista:

Protista is classified into three subdivisions-

i. Protozoans.

ii. Algae.

iii. Mould.

Protozoans:

i. They are unicellular organisms.

ii. They are heterotrophic.

iii. They are eukaryotic.

Algae:

i. Algae are chlorophyll bearing plants. They are generally found in aquatic habitats.

ii. They are capable of photosynthesis.

iii. Algae are unicellular and eukaryotic.

iv. They are autotrophic.

Mould:

i. Mould is a group of fungi.

ii. They are visible to the naked eye.

iii. Moulds are saprophytic organisms.

Classifications of Fungi:

Fungi are classified into five major groups-

i. Chytridiomycota.

ii. Zygomycota.

iii. Ascomycota.

iv. Basidiomycota.

v. Deutromycota.

Chytridiomycota:

i. They are unicellular or multicellular.

ii. Cell wall is composed of chitin and cellulose.

iii. They are aquatic fungi.

Zygomycota:

i. Zygomycotaare haploid.

ii. They reproduce by sexually or asexually.

iii. Asexual reproduction takes place by aplanospores.

iv. Sexual reproduction takes place by gametangial copulation.

Ascomycota:

i. Ascomycotinaare called sac fungi.

ii. cell wall is composed of chitin and glucans.

iii. Sexual reproduction takes place by gametangial copulation, gametangial contact or spermatization.

iv. Asexual reproduction takes place by conidia, oidia, and chylamydospores.

Basidiomycota:

i. Basidiomycotaare composed of chitin.

ii. They are unicellular or multicellular

Deutromycetes:

i. Deutromycetes are also known as fungi imperfecti.

ii. Sexual reproduction is unknown.

iii. They are saprophytes or parasites.

iv. They reproduce by asexual methods.

Lichens:

Lichens are consisting of algae and fungi. Algal partners manufactures the food and the food is absorbed by fungal partners.

Lichens are found in branch or rock surfaces. The fungal partners is known as the mycobiont. The autotroph is known as photobiont.

Economic Importance:

i. Lichens are used as food.

ii. Lichens are used in the treatment of jaundice, diarrhoea, fever, epilepsy, and skin disease.

ii. Litmus is obtained from Rocellatinctoria, R. montagnei.

Viruses:

In 1900, it was generally accepted that many of the recognized human diseases were caused by microorganisms, the first evidence of viruses as causative agents of disease came in 1892 when Ivanowski showed that cell free extracts of diseased tobacco leaves passed through bacteria proof filters, could cause disease in healthy plants. The word virus come from a latin word simply meaning slimy fluid. The first human virus described was the agent which causes yellow fever. The virus was discovered and reported in 1901 by the US Army physician Walter Reed. Beijerinck (1897) coined the Latin name virus meaning poison. Viruses are very small- smaller than the smallest cell. The scientific study of viruses are called virology.

Viruses were first discovered after the development of a porcelain filter called the chamber-land-pasteur filter, which could remove all bacteria visible in the microscope from any liquid sample. Thus virology (study of viruses) is a significant part of microbiology. Viruses cause many more illnesses disease.

General properties of viruses

The virus is made up of a genetic information molecule and a protein layer that protects that information molecule. The core of the virus is made up of nucleic acids, which then make up the genetic information in the form of DNA or RNA. Viruses are much smaller than prokaryotic or eukaryotic cells. Some viruses (bacteriophage) are infect prokaryotic cell or eukaryotic cells. Viruses are non cellular, self replicating agents. They are transmitted very easily from one organism to another organism.

The Structure of Viruses

Viruses come in amazing variety of size and shapes. They are very small and are measured in nanometers, which is one-billionth of a meter. Viruses are seen with a scanning electron microscope. A simple virus particle often called a virion. Virions range in size from about 10 to 400 nm in diameter. Viruses contain either DNA or RNA but not both. Each virion contains only one molecule of nucleic acid, called genome. The protein coat surrounding the genome is called capsid. The capsid is made up of large number of protein subunit called capsomeres.

Symmetry: The capsid is symmetrically arranged around the central nucleic acid. Capsids protect genome from atmosphere. Viral capsids are made of repeated protein subunits. There are three types of capsids symmetry:

i. Cubical (icosahedr

ii. Helical Capsids.

iii. Complex capsids.

i. Cubical (icosahedral) capsids: These viruses appear spherical in shape. The icosahedral made up of equilateral triangles. They have a polygon with 20 sides (facets) and 30 edges. They usally made up of five or six pentamers. Icosahedral capsids contain both pentamers and hexamers. Simian virus 40 (SV 40) onlypentamers.

ii. Helical capsids: Helical capsids are shaped like hollow tube or central cavity that is made by proteins arranged in a circular fashion, creating a disc like shape. They are usally 15-19 nm wide. e.g Tobacco Mosaic virus (TMV), Influenza virus etc

iii. Complex capsids: These virus structures have a combination of icosahedral and helical shape. The head of the virus has an icosahedral shaped within a helical shaped tail. e.g Pox virus and bacteriophages like T_2, T_4 and T_6.

Viriod

Theodor O. Diener discovered a cell-invading plant pathogen 80 times smaller than a viruses; the viriod. Viriods are unique plant pathogens consisting of low molecular weight, autonomously replicated single stranded RNA molecules (approximately 246 to 425 nucleotides (nt) without any functional open reading frame

(ORF) in their genome. The first viriod, potato spindle tuber viriod (PSTVd), was discovered in the late 1960s – early 1970s (Diener, 1971). Viriods are classified into two families : the pospiviroidae, which replicate in the nucleus, and the Avsunviroidae, which replicate in the chloroplast (Tsagris, 2008). They do not code for protein. The viriods potato spindle tuber viriod (PSTVd) and Citrus exocortisviriod (CEVd) are both replicated in the nucleus.

Prion

The word itself derives from proteinaceous infectious particle meaning that the infectious agent consists only of protein with no nucleic acid genome. The term prion refers to abnormal pathogenic that are transmissible and are able to induce abnormal folding of specific normal cellular proteins called prion proteins. Prions cause a variety of neurogenerative diseases in humans and animals. Prions are responsible for bovine spongiform encephalopathy (BSE or mad cow disease), Scarpie (goat and sheep), chronic wasting disease (deer and elk), and the human diseases kuru, fatal, sporadic insomnia, Gerstmann-Strausster-Scheinzer disease, Creutzfeldt-Jacob disease (CJD). In 1950s, disease known as kuru, in Papua New Guinea discovered; neuropathological similarity between kuru, CJD, and scrapie noted. Research suggests that scrapie is caused by abnormal form of a cellular protein. The abnormal form is called PrP^{sc} (for scrapie-associated prion protein) and the normal cellular form is called PrP^{c}.

Chapter 3

Biodiversity and It's Conservation

Biodiversity Patterns:

Darwin noticed three patterns of biological diversity-

i. Species vary globally.

ii. Species vary locally.

and iii. Species vary over time.

Ecologists have studied various patterns of species biodiversity-

i. Latitudinal Gradients- When, we move from equators, towards the poles, the diversity of species is decreases.

ii. Species-Area Realationships: Species-Area Relationships are the number of species found on the Earth.

Log scale represents the following equations-

$\log S = \log C + z \log A$

Here, S= species richness.

A= Area

Z= Slope of the line.

C= Y- intercept

Importance of Biodiversity:

Biodiversity provides us to oxygen, air, water, timber, and many other thing. Biodiversity provides large number of plant species. Biodiversity provides food, medicine, and drugs.

Loss of Biodiversity:

Biodiversity loss is the extinction of loss of species in a certain habitat.

i. Deforestation: We cut down the trees for various purpouses. We destroy the ecosystems.

ii. Overexploitation: Hunting and poaching of the species is the major causes for the loss of biodiversity.

iii. Pollution: Pollution is the major causes for the loss of biodiversity. We know that, plastics are dumped into the ocean surface. Plastics are polluted our environment and earth ecosystems.

Burnning of fossils fuels are polluted into the atmosphere.

iv. Lastly say, climate change and global warming is the major causes for the loss of biodiversity.

Biodiversity Conservation:

Biodiversity conservation is the protection, conservation of biodiversity for sustainable development.

Types of Conservation:

In Situ Conservation:

In situ conservation is the conservation of plant or animal species. In situ conservation includes National Park and sanctuaries, biosphere reserves, nature reserves, reserve and protected forests, preservation plots, reserved forests etc.

Ex Situ Conservation:

Ex situ conservation is the conservation of living organisms. Ex situ conservation includes zoological and botanical parks.

Hotspots:

Biodiversity hotspots are the contain at least 1500 species of vascular plants nowhere else on Earth. Biodiversity includes both flora and fauna. Biodiversity hotspots regions are particularly rich in endemic, rare, and threatened species.

Biodiversity hotspots concept was first introduced by Norman Myers in 1988.

There are at least 34 biodiversity hotspots regions around the World-

Africa

1. Eastern Afro-Montane

2. The Guinean forests of Western Africa

3. Horn of Africa

4. Madagascar and the Indian Ocean Islands

5. Maputoland, Podoland, Albany hotspot

6. Succulent Karou

7. East Malanesian islands

8. South Africa's Cape floristic hotspot

9. Coastal forests of Eastern Africa

Terrestrial Biomes of the World

Asia and Australia

1. Himalayan hotspot

2. The Eastern Himalayas

3. Japan biodiversity hotspot

4. Mountains of South-West China

5. New Caledonia

6. New Zealand biodiversity hotspot

7. Philippine biodiversity hotspot

8. Western Sunda (Indonesia, Malas and Brunei)

9. Wallace (Eastern Indonesia)

10. The Western Ghats of India and Islands of Sri Lanka

11. Polynesia and Micronesian Islands Complex including Hawaii

12. South-Western Australia

North and Central America

1. California Floristic Province

2. Caribbean islands hotspot

3. Modrean pine-oak wood lands of the USA and Mexico border

4. The Mesoamerican forests

Aquatic Biomes of the World

South America

1. Brazil's Cerrado

2. Chilean winter rainfall (Valdivian) Forests

3. Tumbes-Choco-Magdalena

4. Tropical Andes

5. Atlantic forest

Europe and Central Asia

1. Caucasus region

2. Iran-Anatolia region

3. The Mediterranean basin and its Eastern Coastal region

4. Mountains of Central Asia

Endangered Species:

Endangered species is a species i.e very extinct in near future.

Some endangered species are listed below-

i. Orangutan (Pongopygmaeus).

ii. Tasmanian devil (Sarcophilusharrisii).

iii. Gorilla (Gorilla beringei).

iv. Snow leopard (Panthera uncial).

v. Sea otter (Enhydralutris).

vi. Asian elephant (Elephasmaximus).

vii. Blue whale (Balaenopteramusculus).

viii. Whooping Crane (Grusamericana).

ix. Tiger (Panthera Tigris).

x. Giant Panda (Ailuropodamelanoleuca).

Red Data Book:

The Red data book is a public document which is originate from Russia. The Red data book is recording endangered species of plants, animals, and fungi.

The Red data book contains various colour represents coded information sheets i.e listed below-

Red- endangered.

Black- Species, which are confirmed to be extinct.

Amber- Vulnerable.

White- Rare.

Green- endangered but their numbers have started to recover.

Grey- Vulnerable, endangered, or rare.

Biosphere Reserves:

Biosphere reserves are nominated by national government. Biosphere reserves consists of three zones i.e core area, buffer zone, and transition area.

According to UNESCO, Biosphere reserves are areas comprising terrestrial, marine, and coastal ecosystems that are recognised by it's the Man and the Biosphere Programme (MAB, 1971).

List of Biosphere Reserves in India-

i. Nilgiri Biosphere Reserves, Tamil Nadu, Kerala, and Karnataka.

ii. Gulf of Mannar, Tamil Nadu.

iii. Sundarbans Biosphere Reserves, West Bengal.

iv. Nanda Devi Biosphere Reserves, Uttarakhand.

v. Nokrek Biosphere Reserves, Meghalaya.

vi. Pachmarhi Biosphere Reserves, Madhya Pradesh.

vii. Simlipal Biosphere Reserves, Odisha.

viii. Great Nicobar Biosphere Reserves, Andaman and Nicobar Islands.

ix. Agasthyamalai Biosphere Reserves, Kerala and Tamil Nadu.

National Park:

National parks are recognized by national government. National parks are protect the flora and fauna. National parks is an area where reserved of the wildlife and biodiversity.

The first national park is Yellowstone National Park in 1872. The largest national park is Hemis National Park, Jammu and Kashmir.

List of National park in India:

i. Campbell Bay National Park, Andaman and Nicobar Islands.

ii. Rajiv Gandhi National Park, Andhra Pradesh.

iii. Namdapha National Park, Arunachal Pradesh.

Iv, Kaziranga National Park, Assam.

v. Valmiki National Park, Bihar.

vi. Guru Ghasidas National Park, Chattisgarh.

vii. Mollem National Park, Goa.

viii. Vansda National Park, Gujarat.

ix. Kalesar National Park, Haryana.

x. Pin Valley National Park, Himachal Pradesh.

xi. Dachigam National Park, Jammu and Kashmir.

xii. Betla National Park, Jharkhand.

xiii. Bandipur National Park, Karnataka.

xiv. Silent Valley National Park, Kerala.

xv. Kanha National Park, Madhya Pradesh.

xvi. Pench National Park, Maharashtra.

xvii. Keibul-Lamjao National Park, Manipur.

xviii. Nokrek Ridge National Park, Meghalaya.

xx. Buxa National Park, West Bengal.

Sanctuaries:

Sanctuaries are an area where animals are protected from hunting.

List of Wildlife Sanctuaries in India:

i. Ariak Island, Andaman and Nicobar Island.

ii. Barren Island, Andaman and Nicobar Island.

iii. Kolleru Bird Sanctuary, Andhra Pradesh.

iv. Pulicat Lake Bird Sanctuary, Andhra Pradesh.

v. Pakke Tiger Reserve, Arunachal Pradesh.

vi. Bornadi Wildlife Sanctuary, Assam.

vii. Bhimbandh Wildlife Sanctuary, Bihar.

Viii. Udaypur Wildlife Sanctuary, Bihar.

ix. Achanakmar Wildlife Sanctuary, Chattisgarh.

x. Dadra and Nagar Haveli Wildlife Sanctuary, Dadra Nagar Haveli.

xi. Salim Ali Bird Sanctuary, Goa.

xii. Bhindawas Wildlife Sanctuary, Haryana.

xiii. Gir Wildlife Sanctuary, Gujarat.

xiv. Dalma Wildlife Sanctuary, Jharkhand.

xv. Neyyar Wildlife Sanctuary, Kerala.

xvi. Great Indian Bustard Sanctuary, Maharastra.

xvii. Kotgarh Wildlife Sanctuary, Odisha.

xviii. Maenam Wildlife Sanctuary, Sikkim.

xx. kumbhalgarh Wildlife Sanctuary, Rajasthan.

xxi. Vedanthangal Bird Sanctuary, Tamil Nadu.

xxii. Pocharam Wildlife Sanctuary, Telangana.

xxiii. Gumti Wildlife Sanctuary, Tripura.

xxiv. Hastinapur Wildlife Sanctuary, Uttar Pradesh.

xxv. Binsar Wildlife Sanctuary, Uttarakhand.

xxvi. Chapramari Wildlife Sanctuary, West Bengal.

Ramsar Sites:

The Convention on wetlands are also known as Ramsar Convention. Ramsar Convention was established in 1971. Ramsar is a city of Iran.

International Organizations Partners:

There are six organizations are listed below-

- Birdlife International.
- International Union for Conservation of Nature (IUCN).
- International Water Management Institute (IWMI).
- Wetlands International.
- WWF International.
- Wildfowl and wetlands trust (WWT).

List of Ramsar Sites-

i. Ashtamudi wetland, Kerala.

ii. Bhitarkanika Mangroves, Odisha.

iii. Bhoj wetland, Madhya Pradesh.

iv. Chandra Taal, Himachal Pradesh.

v. Chilka Lake, Odisha.

vi. DeeporBeel, Assam.

vii. East Kolkata Wetlands, West Bengal.

viii. Harike Wetlands, Punjab.

ix. Hokera Wetland, Jammu and Kashmir.

x. Kanjli Wetland, Punjab.

xi. Kolleru Lake, Andhra Pradesh.

xii. Loktak Lake, Manipur.

xiii. Pong Dam Lake, Himachal Pradesh.

References:

1. https://byjus.com/biology/what-is-living/

2. https://www.theguardian.com/news/2018/mar/12/what-is-biodiversity-and-why-does-it-matter-to-us

3. https://www.nationalgeographic.org/encyclopedia/biodiversity/

4. https://www.vedantu.com/biology/biodiversity

5. https://en.m.wikipedia.org/wiki/Biodiversity

6. https://www.safeopedia.com/definition/2992/species-diversity

7. https://socratic.org/questions/what-is-ecosystem-diversity

8. http://www.coastalwiki.org/wiki/Ecosystem_diversity

9. https://www.cbd.int/gti/taxonomy.shtml

10. https://byjus.com/biology/concept-of-species/

11. https://byjus.com/biology/binomial-nomenclature/

12. https://en.m.wikipedia.org/wiki/Botanical_garden

13. http://www.biologydiscussion.com/articles/importance-of-botanical-gardens/6522

14. https://www.padeepz.net/herbaria-and-uses-importance-of-herbaria/

15. https://en.m.wikipedia.org/wiki/Museum

16. https://www.takshilalearning.com/class-12-biology-patterns-of-biodiversity/

17. https://soe.environment.gov.au/theme/biodiversity/topic/2016/importance-biodiversity

18. http://ete.cet.edu/gcc/?/biodiversity_importance/

19. https://openoregon.pressbooks.pub/envirobiology/chapter/123/

20. https://mashable.com/2015/05/23/biodiversity-threats/

21. https://www.greenfacts.org/en/biodiversity/l-3/4-causes-desertification.htm

22. https://byjus.com/biology/biodiversity-conservation/

23. https://swww.cepf.net/our-work/biodiversity-hotspots/hotspots-defined

24. https://m.jagranjosh.com/general-knowledge/biodiversity-hotspots-of-the-world-1523356211-1

25. https://www.britannica.com/list/10-of-the-most-famous-endangered-species

26. https://www.unescomedcenter.org/en/biosphere-reserves

27. https://en.m.wikipedia.org/wiki/Biosphere_reserves_of_India

28. https://simple.m.wikipedia.org/wiki/National_park

29. https://www.careerpower.in/national-parks-india.html

30. https://en.m.wikipedia.org/wiki/List_of_wildlife_sanctuaries_of_India

31. https://en.m.wikipedia.org/wiki/Ramsar_Convention

32. https://en.m.wikipedia.org/wiki/List_of_Ramsar_sites_in_India

33. https://en.m.wikipedia.org/wiki/Species%E2%80%93area_relationship

1. https://doi.org/10.1016/j.mrgentox.2008.09.017

2. Shukla S, Mahata S, Shishodia G, Pandey A, Tyagi A, Vishnoi K, et al. (2013) Functional Regulatory Role of STAT3 in HPV16-Mediated Cervical Carcinogenesis. PLoS ONE 8(7): e67849. doi:10.1371/journal.pone.0067849

3. https://doi.org/10.1111/j.1749-6632.2009.04911.x

4. Cytokine Growth Factor Rev. 2016 October ; 31: 1–15. doi:10.1016/j.cytogfr.2016.05.001.

5. doi: http://dx.doi.org/10.1101/137653.

6. https://doi.org/10.3109/10799891003786218

7. https://doi.org/10.1186/1476-4598-9-282

8. Mol Cell Biochem. 2009 Oct;330(1-2):193-9. doi: 10.1007/s11010-009-0133-2. Epub 2009 May

9. ClinSci (Lond). 2006 May;110(5):525-41.

10. DayaluNaik SL, Kumar V, Joshi R, Bharti AC. HPV Inflammation Mediate IL-6 through STAT3 Signaling Pathway in Different Grades of Cervical Cancer. J Cancer Res Molecul Med. 2016;3(1): 103.

11. Ranbir C. Sobti, Neha Singh, ShowketHussain, VanitaSuri, MausumiBharadwaj&Bhudev C. Das (2010) Deregulation of STAT-5 isoforms in the development of HPV-mediated cervical carcinogenesis, Journal of Receptors and Signal Transduction, 30:3, 178-188, DOI: 10.3109/10799891003786218

12. Current cancer drug targets · March 2007 DOI: 10.2174/156800907780006869 · Source: PubMed

13. Madigan, Michael T, John M. Martinko, and Jack Parker. Brock Biology of Microorganisms. Upper Saddle River, NJ: Prentice Hall/Pearson Education, 200

1. Bacteriorhodopsin and Related Pigments of Halobacteria. https://www.ncbi.nlm.nih.gov/pubmed/6287921.

2. Eme, L. & Doolittle, W. F. (2015). Archaea.CurrBiol 25, R845–R875.Diversity and Abundance of Korarchaeota in Terrestrial Hot ... https://www.nature.com/articles/ismej2009126.

3. Spang, A. & others (2015). Complex archaea that bridge the gap between prokaryotes and eukaryotes. Nature 521, 173–179.

4. Williams, T. A. & others (2013). An archaeal origin of eukaryotes supports only two primary domains of life. Nature 504, 231–236.

5. Woese C. R., Kandler O. &Wheelis M. L. (1990). Towards a natural system of organisms: proposal for the domains Archaea, Bacteria, and Eucarya. ProcNatlAcadSci USA 87, 4576–4579.

6. T. D. Brock, M. T. Madigan, J. M. Martinko, & J. Parker. 1994. Biology of Microorganisms, 7th ed. (New Jersey: Prentice Hall).

7. W. Ford Doolittle. 1992. What are the archaebacteria and why are they important? Biochemical Society Symposium 58: 1-6.

8. G. E. Fox, L. J. Magrum, W. E. Balch, R. S. Wolfe, & C. R. Woese, 1977. Classification of methanogenic bacteria by 16S ribosomal RNA characterization. Proc. Natl. Acad. Sci. USA 74: 4537-4541.

9. McInerney, J. O., M. Wilkinson, J. W. Patching, T. M. Embley and R. Powell. 1995. Recovery and phylogenetic analysis of novel archaealrRNA sequences from a deep-sea deposit feeder. Appl. Envir. Microbiol. 61: 1646-1648.

10. Rieger, G., R. Rachel, R Hermann and K.O. Stetter. 1995. J. Struct. Biol. 115:78-87.

11. K. Horikoshi& W. D. Grant (eds.). 1998. Extremophiles -- Microbial Life in Extreme Environments (New York: Plenum).

12. John L. Howland. 2000. The Surprising Archaea (New York & Oxford: Oxford University Press). M. T. Madigan & B. L. Marrs, 1997.Extremophiles. Scientific American (Apr): 82-87. C. R. Woese, 1981. Archaebacteria. Scientific American (Jun): 98-122.

13. C. R. Woese& G. E. Fox, 1977. Phylogenetic structure of the prokaryotic domain: The primary kingdoms. Proc. Natl. Acad. Sci. USA 74: 5088-5090.

14. Bintrim, S.B., Donohue, T.J., Handelsman, J., Roberts, G.P., Goodman, R.M. 1997. Molecular phylogeny of Archaea from soil. Proc. Natl. Acad. Sci. USA 94:277-282.

15. Bloechl, E., R. Rachel, S. Burggraf, D. Hafenbradl, H.W. Jannasch and K.O. Stetter. 1997. Extremophiles 1:14-21.

16. Brock, T. D. 1978. Thermophilic microorganisms and life at high temperatures.New York, Springer-Verlag.

17. Fuhrman, J. A., K. McCallum and A. A. Davis. 1992. Novel marine archaebacterial group from marine plankton. Nature 356: 148-149.

18. Hershberger, K.L., S. M. Barns, A.-L. Reysenbach, S.C. Dawson and N.R. Pace. 1996. Wide diversity of Crenarchaeota. Nature 384:420.

19. Kjems, J., N. Larsen, J. S. Dalgaard, R. A. Garrett and K. O. Stetter. 1992. Phylogenetic relationships amongst the hyperthermophilicArchaea determined from partial 23S rRNA gene sequences. System. Appl. Microbiol. 15: 203-208.

20. Segerer, A. H., S. Burggraf, G. Fiala, G. Huber, R. Huber, U. Pley and K. O. Stetter. 1993. Life in hot springs and hydrothermal vents. Orig. Life Evol.Biosph. 23: 77-90.

21. Stetter, K. O. 1996. Hyperthermophilicprocaryotes.FEMS Microbiol.Revs. 18: 149-158. Voelkl, P., R. Huber, E. Drobner, R. Rachel, S. Burggraf, A. Trincone and K.O. Stetter. 1993. Appl. Environ. Microbiol. 59:2918-2926.

1. https://animaldiversity.org/accounts/Animalia/

2. https://en.m.wikipedia.org/wiki/Animal

3. https://www.siyavula.com/read/science/grade-10-lifesciences/biodiversity-and-classification/09-biodiversity-and-classification-04

4. https://byjus.com/biology/monera/

5. https://ucmp.berkeley.edu/bacteria/cyanointro.html

6. http://www.beachapedia.org/Cyanobacteria

7. https://byjus.com/biology/protista/

8. https://en.m.wikipedia.org/wiki/Protist

9. https://www-livescience-com.cdn.ampproject.org/v/s/www.livescience.com/amp/54979-what-are-algae.html?amp_js_v=a3&amp_gsa=1&usqp=mq331AQFKAGwASA%3D#aoh=15889070620293&amp_ct=1588907071626&csi=1&referrer=https%3A%2F%2Fwww.google.com&amp_tf=From%20%251%24s&ampshare=https%3A%2F%2Fwww.livescience.com%2F54979-what-are-algae.html

10. https://www.britannica.com/science/algae/Physical-and-ecological-features-of-algae

11. https://eol.org/docs/discover/algae

12. https://courses.lumenlearning.com/biology2xmaster/chapter/classification-of-fungi/

13. https://en.m.wikipedia.org/wiki/Chytridiomycota

14. http://website.nbm-mnb.ca/mycologywebpages/NaturalHistoryOfFungi/Chytridiomycota.html

15. Alexopoulos, C. J., C. W. Mims and M. Blackwell. 1996. Introductory mycology. John Wiley and Sons, New York.

Benny, G. L., R. A. Humber and J. B. Morton. 2001. Zygomycota: Zygomycetes. Pp. 113-146. In: The Mycota VII. Systematics and Evolution. Part A. (McLaughlin, D. J., McLaughlin, E. G. and Lemke, P. A., eds.). Springer-Verlag, New York.

Benny G. L. and K. O'Donnell. 2000. Amoebidiumparasiticum is a protozoan, not a Trichomycete. Mycologia 92: 1133-1137.

Berbee, M. L. and J. W. Taylor. 2001. Fungal molecular evolution: gene trees and geologic time. Pp. 229-245. In: The Mycota VII. Systematics and Evolution. Part B. (McLaughlin, D. J., McLaughlin, E. G. and Lemke, P. A., eds.). Springer-Verlag, New York.

Berbee, M. L. and J. W. Taylor. 1993. Dating the evolutionary radiations of the true fungi. Can. J. Bot. 71: 1114-1127.

Bidochka, M. J., S. R. A. Walsh, M. E. Ramos, R. J. St. Leger, J. C. Silver and D. W. Roberts. 1996. Fate of biological control introductions: monitoring an Australian fungal pathogen of grasshoppers in North America. Proc. Natl. Acad. Sci. USA 93: 918-921.

Blackwell, M., and D. Malloch. 1989. Similarity of Amphoromorpha and secondary capilliconidia of Basidiobolus. Mycologia 81: 735-741.

Bruns, T. D., R. Vilgalys, S. M. Barns, D. Gonzalez, D. S. Hibbett, D. J. Lane, L. Simon, S. Stickel, T. M. Szaro, W. G. Weisburg and M. L. Sogin. 1992. Evolutionary relationships within the fungi: analyses of nuclear small subunit RNA sequences. Mol. Phylogenet. Evol. 1: 231-241.

Cavalier-Smith, T. 1998.A revised six-kingdom system of life.Biol. Rev. 73: 203-266.

Cole, G. T. and R. A. Samson. 1979. Patterns of development in conidial fungi. Pitman, London.

deHoog, G. S., J. Guarro, J. Gene and M. J. Figueras. 2000. Atlas of clinical fungi, second addition. CentraalbureauvoorSchimmelcultures, Baarn and Delft, The Netherlands.

Eslava, A. P., M. I. Alvarez, and M. Delbrück. 1975. Meiosis in Phycomyces. Proc. Natl. Acad. Sci. USA 72: 4076-4080.

Hajek, A. E. 1999. Pathology and epizootiology of Entomophagamaimaga infections in forest Lepidoptera.Microbiol.Mol. Biol. Rev. 63: 814-835.

Heckman, D. S., D. M. Geiser, B. R. Eidell, R. L. Stauffer, N. L. Kardos and S. B. Hedges. 2001. Molecular evidence for the early colonization of land by fungi and plants. Science 293: 1129-1133.

Hesseltine, C. W. 1991. Zygomycetes in food fermentations. The Mycologist 5: 162-169.

Hibbett, D. S., M. Binder, J. F. Bischoff, M. Blackwell, P. F. Cannon, O. E. Eriksson, S. Huhndorf, T. James, P. M. Kirk, R. Lucking, H. T. Lumbsch, F. Lutzoni, P. B. Matheny, D. J. McLaughlin, M. J. Powell, S. Redhead, C. L. Schoch, J. W. Spatafora, J. A. Stalpers, R. Vilgalys, M. C. Aime, A. Aptroot, R. Bauer, D. Begerow, G. L. Benny, L. A. Castlebury, P. W. Crous, Y.-C. Dai, W. Gams, D. M. Geiser, G. W. Griffith, C. Gueidan, D. L. Hawksworth, G. Hestmark, K. Hosaka, R. A. Humber, K. D. Hyde, J. E. Ironside, U. Koljalg, C. P. Kurtzman, K.-H. Larsson, R. Lichtwardt, J. Longcore, J. Miadlikowska, A. Miller, J.-M.Moncalvo, S. Mozley-Standridge, F. Oberwinkler, E. Parmasto, V. Reeb, J. D. Rogers, C. Roux, L. Ryvarden, J. P.

Sampaio, A. Schüßler, J. Sugiyama, R. G. Thorn, L. Tibell, W. A. Untereiner, C. Walker, Z. Wang, A. Weir, M. Weiss, M. M. White, K. Winka, Y.-J. Yao and N. Zhang. 2007. A higher-level phylogenetic classification of the Fungi. Mycol. Res. 111: 509-547.

James, T. Y., D. Porter, C. A. Leander, R. Vilgalys and J. E. Longcore. 2000. Molecular phylogenetics of the Chytridiomycota supports the utility of ultrastructural data in chytrid systematics. Can. J. Bot. 78: 336-350.

James, T. Y., F. Kauff, C. Schoch, P. B. Matheny, V. Hofstetter, C. Cox, G. Celio, C. Gueidan, E. Fraker, J. Miadlikowska, H. T. Lumbsch, A. Rauhut, V. Reeb, A. E. Arnold, A. Amtoft, J. E. Stajich, K. Hosaka, G.-H. Sung, D. Johnson, B. O'Rourke, M. Crockett, M. Binder, J. M. Curtis, J. C. Slot, Z. Wang, A. W. Wilson, A. Schüßler, J. E. Longcore, K. O'Donnell, S. Mozley-Standridge, D. Porter, P. M. Letcher, M. J. Powell, J. W. Taylor, M. M. White, G. W. Griffith, D. R. Davies, R. A. Humber, J. B. Morton, J. Sugiyama, A. Y. Rossman, J. D. Rogers, D. H. Pfister, D. Hewitt, K. Hansen, S. Hambleton, R. A. Shoemaker, J. Kohlmeyer, B. Volkmann-Kohlmeyer, R. A. Spotts, M. Serdani, P. W. Crous, K. W. Hughes, K. Matsuura, E. Langer, G. Langer, W. A. Untereiner, R. Lücking, B. Büdel, D. M. Geiser, A. Aptroot, P. Diederich, I. Schmitt, M. Schultz, R. Yahr, D. Hibbett, F Lutzoni, D. McLaughlin, J. Spatafora, and R. Vilgalys. 2006a. Reconstructing the early evolution of the fungi using a six gene phylogeny. Nature 443:818-822.

Jensen, A. B., A. Gargas, J. Eilenberg and S. Rosendahl. 1998. Relationships of the insect-pathogenic order Entomophthorales (Zygomycota, Fungi) based on phylogenetic analyses of nuclear small subunit ribosomal DNA sequences (SSU rDNA). Fungal Genet. Biol. 24: 325-334.

Keeling, P. J. 2003.Congruent evidence from alpha-tubulin and beta-tubulin gene phylogenies for a zygomycete origin of microsporidia.Fungal Genet. Biol. 38: 298-309.

Keeling, P. J., M. A. Luker and J. D. Palmer. 2000. Evidence from beta-tubulin phylogeny that microsporidia evolved from within the Fungi. Mol. Biol. Evol. 17: 23-31.

Kirk, P. M., P. F. Cannon, J. C. David, and J. Stalpers. 2001. Ainsworth and Bisby's Dictionary of the Fungi. 9th ed. CAB International, Wallingford, UK.

Krejzova, R. 1978. Taxonomy, morphology and surface structure of Basidiobolus sp. isolate. J. Invertebr. Pathol. 31: 157-163.

Aime, M. C., P. B. Matheny, D. A. Henk, E. M. Frieders, R. H. Nilsson, D. J. McLaughlin, L. J. Szabo, and D. S. Hibbett. 2006. An overview of the higher-level classification of Pucciniomycotina based on combined analyses of nuclear large and small subunit rDNA sequences. Mycologia 98: 869-905.

Alexopoulos, C.J., Mims, C.W. and Blackwell, M. 1996. Introductory Mycology. John Wiley and Sons, New York.

Arora, D. 1986. Mushrooms Demystified. Ten Speed Press, Berkeley, California.

Bauer, R., Begerow, D., Oberwinkler, F., Piepenbring, M. and Berbee, M. L 2001.Ustilaginomycetes. Pp. 57-84. In: The Mycota VII. Systematics and Evolution. Part B. (Mclaughlin, D. J., McLaughlin, E. G. and Lemke, P. A., eds.). Springer-Verlag, Berlin.

Begerow D, Stoll M, and R. Bauer. 2006. A phylogenetic hypothesis of Ustilaginomycotina based on multiple gene analyses and morphological data. Mycologia 98: 906–916.

Benjamin, D.R. 1995. Mushrooms: poisons and panaceas. W.H. Freeman and Company, New York.

Buller, A.H.R. 1909-1934. Researches on Fungi (6 vols.)Longmans, Green and Co., London.

Fell, J. W., Boekhout, T., Fonseca, A. and Sampaio J.P. 2001. Basidiomycetous yeasts. Pp. 1-36. In: The Mycota VII.

Systematics and Evolution. Part B. (Mclaughlin, D. J., McLaughlin, E. G. and Lemke, P. A., eds.). Springer-Verlag, Berlin.

Hibbett, D. S. 2006. A Phylogenetic overview of the Agaricomycotina.Mycologia 98: 917-925.

Hibbett, D. S., and Binder, M. 2001. Evolution of marine mushrooms. Biol. Bull. 201:319-322.

Hibbett, D. S., M. Binder, J. F. Bischoff, M. Blackwell, P. F. Cannon, O. E. Eriksson, S. Huhndorf, T. James, P. M. Kirk, R. Lücking, T. Lumbsch, F. Lutzoni, P. B. Matheny, D. J. Mclaughlin, M. J. Powell, S. Redhead, C. L. Schoch, J. W. Spatafora, J. A. Stalpers, R. Vilgalys, M. C. Aime, A. Aptroot, R. Bauer, D. Begerow, G. L. Benny, L. A. Castlebury, P. W. Crous, Y.-C. Dai, W. Gams, D. M. Geiser, G. W. Griffith, C. Gueidan, D. L. Hawksworth, G. Hestmark, K. Hosaka, R. A. Humber, K. Hyde, J. E. Ironside, U. Kõljalg, C. P. Kurtzman, K.-H. Larsson, R. Lichtwardt, J. Longcore, J. Miądlikowska, A. Miller, J.-M.Moncalvo, S. Mozley-Standridge, F. Oberwinkler, E. Parmasto, V. Reeb, J. D. Rogers, C. Roux, L. Ryvarden, J. P. Sampaio, A. Schüßler, J. Sugiyama, R. G. Thorn, L. Tibell, W. A. Untereiner, C. Walker, Z. Wang, A. Weir, M. Weiß, M. M. White, K. Winka, Y.-J. Yao, and N. Zhang. 2007. A higher-level phylogenetic classification of the Fungi. Mycological Research 111: 509-547.

Hibbett, D. S. and Thorn, R. G. 2001.Homobasidiomycetes. Pp. 121-170. In: The Mycota VII. Systematics and Evolution. Part B. (Mclaughlin, D. J., McLaughlin, E. G. and Lemke, P. A., eds.). Springer-Verlag, Berlin.

Ingold, C.T. 1939. Spore discharge in land plants. Oxford University Press, London, UK.

Ingold, C.T. 1991.A view of the active basidium in heterobasidiomycetes. Mycol. Res. 95: 618-621.

Kirk, P.M., Cannon, P.F., David, J.C., and Stalpers, J. 2001.Ainsworth and Bisby's Dictionary of the Fungi.9th ed. CAB International, Wallingford, UK.

Kohlmeyer, J., and Kohlmeyer, E. 1979. Marine Mycology—The Higher Fungi. Academic Press, New York.

McLaughlin, D.J. 1982. Ultrastructure and cytochemistry of basidial and basidiospore development. In: Basidium and Basidiocarp. Wells, K. and Wells, E.K. (eds.) Springer-Verlag, New York.

McLaughlin, D.M., Beckett, A. and Yoon, K.S. 1985. Ultrastructure and evolution of ballistosporicbasidiospores. Bot. J. Linnean Soc. 91: 253-271.

McLaughlin, D.J., Frieders, E.M. and Lü, Haisheng. 1995. A microscopist's view of heterobasidiomycete phylogeny. Stud. Mycol. 38: 91-109.

Money, N. P. 1998. More g's than the space shuttle: ballistospore discharge. Mycologia 90:547-558.

Mueller, U. G., S. A. Rehner, and T. R. Schultz. 1998. The evolution of agriculture in ants. Science 281: 2034-2038.

Oberwinkler, F. 1977. Das neue System der Basidiomyceten. In: BeiträgezurBiologie der niederenPflanzen. Frey, H., Hurka, H., Oberwinkler, F. (eds.) G. Fischer, Stuttgart.

Oberwinkler, F. 1987. Heterobasidiomycetes with ontogenetic yeast-stages - systematic and phylogenetic aspects. Stud. Mycol. 30: 61-74.

Prillinger, H. Dörfler, C. Laaser, G. and Hauska, G. 1990. A contribution to the systematics and evolution of higher fungi:

yeast-types in the basidiomycetes. Part III: Ustilago-type. Z. Mycol. 56: 251-278.

Prillinger, H., Laaser, G., Dörfler, C. and Ziegler, K. 1991. A contribution to the systematics and evolution of higher fungi: yeast-types in the basidiomycetes. Part IV: Dacrymyces-type, Tremella-type. Sydowia 43: 170-218.

Pringle, A., S. N. Patek, M. Fischer, J. Stolze, and N. P. Money. 2005. The captured launch of a ballistospore. Mycologia 97: 866-871.

Smith, S. E. and Read, D. J. 1997. Mycorrhizal symbiosis. Academic Press, San Diego.

Sugiyama, J., Fukagawa, M, Chiu, S.-W.andKomagata, K. 1985. Cellular carbohydrate composition, DNA base composition, ubiquinone systems, and diazonium blue B color test in the genera Rhodosporidium, Leucosporidium, Rhodotorula and related basidiomycetous yeasts. J. Gen. Appl. Microbiol. 31: 519-550.

Summerbell, R.C. 1985. The staining of filamentous fungi with diazonium blue B. Mycologia 77: 587-593.

Swann, E.C., Frieders, E.M. and McLaughlin, D.J. 1999. Microbotryum, Kriegeria and the changing paradigm in basidiomycete classification. Mycologia 91:51-66.

Swann, E.C., Frieders, E.M. and McLaughlin, D.J. 2001. Urediniomycetes. Pp. 37-56. In: The Mycota VII. Systematics and Evolution. Part B. (Mclaughlin, D. J., McLaughlin, E. G. and Lemke, P. A., eds.). Springer-Verlag, Berlin.

Swann, E.C. and Taylor, J.W. 1993. Higher taxa of basidiomycetes: an 18S rRNA gene perspective. Mycologia 85: 923-936.

Swann, E.C. and Taylor, J.W. 1995. Phylogenetic perspectives on basidiomycete systematics: evidence from the 18S rRNA gene. Canad. J. Bot. 73: S862-S868.

Webster, J., Davey, R.A. and Ingold, C.T. 1984a. Origin of the liquid in Buller's drop. Trans. Br. Mycol. Soc. 83: 524-527.

Webster, J., Davey, R.A., Duller, G.A. and Ingold, C.T. 1984b. Ballistospore discharge in Itersoniliaperplexans. Trans. Br. Mycol. Soc. 82: 13-29.

Wells, K., and Bandoni, R. J. 2001.Heterobasidiomycetes. Pp. 85-120. In: The Mycota VII. Systematics and Evolution. Part B. (Mclaughlin, D. J., McLaughlin, E. G. and Lemke, P. A., eds.). Springer-Verlag, Berlin.

Wheeler, Q. and Blackwell, M. 1984. Fungus-insect relationships. Columbia University Press, New York.

Yoon, K.S. and McLaughlin, D.J. 1986. Basidiosporogenesis in Boletus rubinellus II. Late spore development. Mycologia 78: 185-197.

www.ingramcontent.com/pod-product-compliance
Lightning Source LLC
Chambersburg PA
CBHW040236220526
45473CB00001B/265